SOTHIS PRESS

Sothis Press aims to completely reimagine the public domain book with an emphasis on universal access to captivatingly designed print editions. We believe that technology will enable every single book ever published to be perpetually accessible in beautiful print form. Sothis is beauty in ubiquity.

www.sothis.press
1 Waterhouse Square, London, ECIN 2ST

# Lake Hopatcong the Beautiful

A Plea for Its Dedication as a Public Park and
for Its Preservation as a Pleasure and Health
Resort for the Benefit of All the People

Hudson Maxim

SOTHIS PRESS

ISBN 978-1-0368-0575-3

*Lake Hopatcong the Beautiful*
is in the public domain.

This edition is published by
Sothis Press © 2023. The company
is registered in England and Wales.
The Company Number is 14900389.

# LAKE HOPATCONG THE BEAUTIFUL

# A PLEA

## for its Dedication as a Public Park and for its Preservation as a Pleasure and Health Resort for the Benefit of all the People

BY

HUDSON MAXIM

MAXIM PARK
BOROUGH OF HOPATCONG
LANDING P. O.
NEW JERSEY

THE McCONNELL PRINTING CO.
230-242 WILLIAM STREET, NEW YORK

# PREFACE

This book is a plea for the life of New Jersey's most beautiful lake, and a plea for the life of the communities around that lake, and a plea also for the life of thousands of people in all parts of the State who, if that lake is preserved as a summer and health resort, will continue to find renewal of health and life in visiting it.

This plea is addressed to the Morris Canal Investigation Committee, to the Legislature, and to the people of the State of New Jersey.

Lake Hopatcong is threatened with ruination; the prosperous communities around it are threatened with extinction, and our most precious mountain-lake resort is in imminent danger of being wiped off the map.

Armed with fore-knowledge of the intentions of the enemy, and having the facts, the truth and justice as my allies, I hope and I trust that I shall be able to convince any citizen of this State that he should join issue with the defenders where also his own interests lie, and help to prevent a great wrong being done to the whole population of the State as well as to the immediate lake communities.

People of the State of New Jersey, let it be known to you that there is a movement to take away from you, without any sort of compensation whatsoever, a very valuable lake and summer resort property which belongs to you.

Citizen of the State of New Jersey, you are a member of a company, an actual shareholder in a corporation, which is the State, in which also every man, woman and child in the State are shareholders. A part of your annual dividends, due and payable to you, are your rights and privileges in your lakes, public parks and summer and health resorts.

There is a scheme afoot to rob you of your best and most valuable mountain-lake and summer resort property, in order to appropriate Lake Hopatcong as a city water supply, which would benefit only about five per cent. of the population of the State, at the expense of the rest—in other words it is proposed to rob twenty Peters to pay one Paul, and you and I, reader, are among the Peters.

Now, it is your duty to yourself and to your family, and to every friend and neighbor of yours, that you should lend a helping voice of protest to prevent you and them from being robbed. The reasons why you also should raise your voice in defence of your own and the people's own, are exactly the same reasons which are actuating me to make this plea.

It matters not whether or not my interests may be greater or less than yours—our interests are certainly identical and mutual.

My sense of duty has led me to make this plea. Your sense of duty should, at least, make you read it.

No special activities or efforts are asked of you. If you will read this plea through, your activities and your efforts will take care of themselves.

HUDSON MAXIM.

Landing, New Jersey, Feb. 3, 1913.

# CONTENTS

## Ownership by Seizure Versus Ownership by Purchase

Our legal statutes are coercive rules of behavior based upon our past customs and practices. The customs and practices of an enlightened, mechanical and scientific age are vastly different from those of a primitive, unmechanical and unscientific age. Consequently, many laws which may have been wise and applicable in a simpler state of society can not, with wisdom, be enforced in a higher and more complex state of society.

Robert G. Ingersoll once said, with a good measure of truth, that the world has not been fit for a gentleman to live in for much more than fifty years.

At any rate, it is probable that the world has advanced more during the last hundred and fifty years in all things which make for complete living than it had previously advanced in all the long eons since the world thawed out of the ancient ice.

Consequently, during the nearly hundred years of the existence of the Morris Canal, great changes have taken place in all the mechanism of human society, and the laws which were made to govern conditions three generations ago are in many respects inapplicable to the present generation.

Most of the laws which have been enacted in every age and every generation of men have been to amend or repeal pre-existing laws, which the wisdom of wider experience proved unjust or impracticable, or which were rendered inapplicable under the changed conditions of society.

The need of a law must, therefore, of necessity precede the enactment of the law. The supply of wise laws can never quite meet the demand for those laws.

Statutory laws are an aggregation of fixed rules of conduct, which of necessity can never quite meet the multifarious requirements of infinite variation under continually changing conditions.

The accumulation of laws and precedents now upon our statute books is so massive that the study necessary to the highest Chinese scholarship could scarcely master it.

No wonder then that laws are often enacted which are unconstitutional or which contradict already existing laws.

The many interests, necessities and relations of individuals, communities and corporations connected with the abandonment of the Morris Canal apparently require the enactment of special laws for their adjustment, and the plan for bringing about this desirable result ought to be reached by the exercise of concerted common sense, with evident necessary regard for the eternal fitness of things. It is simply a business men's proposition.

The chief problem for the lawyers to settle in connection with the abandonment of the Morris Canal appears to be the relation of the validity of ownership by purchase to the validity of ownership by seizure.

When the Morris Canal and Banking Company built its dam or water wall at the outlet of Lake Hopatcong, raising the high water level of the Lake eleven feet above its former level, thereby enormously increasing the storage capacity of the Lake, that Company acquired by actual purchase, either of the land itself or of the right of flowage, only a very small part of the total area over which the water was raised. By far the larger part of the lands affected by raising the Lake were never purchased by the Canal Company, and no right whatever to flow them was ever purchased by that Company. A large part of the bed of the old lake and shore lands affected belonged at that time to the East Jersey Board of Proprietors.

About one-half of the bed of the old Lake and of the newly flowed lands were purchased by Nathaniel Niles, from the East Jersey Board of Proprietors, in 1882, and they are still the property of private owners.

These lands have a total area of about one thousand, one hundred and sixty-five acres, and comprise about half of the total area covered by the waters of the Lake; and yet the Canal Company, neither by purchase nor by condemnation, acquired any right whatsoever to flow these lands.

A large amount of land owned by other persons was flowed, with similar lack of recognition of the rights of the owners, no payment or compensation of any kind ever having passed from the Canal Company into the hands of the owners.

We are told that when a public service institution, like a railroad or a canal, has acquired private property by user instead of by purchase, the corporation can not be dispossessed by the actual owner of the property. We have the consolation of being advised, however, that if and when the occupancy and operations of the public service institution cease, then all rights acquired by it revert to the original owner. But even this consolation is lately becoming tainted by certain adverse legal opinions to the effect that property which has been put to one public use, may be diverted to another public use without the consent of the actual owner and without compensation to the owner, provided that the new use is no more onerous or disadvantageous to him than the old use, but this legal opinion lies so close to the margin of constitutionality that the possessors of it are apt to hold it with a tenure somewhat proportionate to the involvement of their personal interests.

Of course such procedure is not justice to the individual property owner. No one pretends or can pretend that it is justice—only that it is the law.

The mistake is often made of confounding law with justice. Law is merely an attempt to attain justice, when it is not enacted by some powerful interest for the purpose of thwarting justice.

Justice is the target at which legislators are supposed to aim. But as legislators are often rather poor legal marksmen, it is rare indeed that a bull's eye is made when justice is aimed at.

It may be said of law, that it is a codification of custom, mainly aiming at justice; but, as is always true of bad marksmanship, there are often many casualties among innocent bystanders.

A small boy, seeing a big piece of apple pie cut presumably for his grandmother, exclaimed with indignation, "Oh, what a great piece of pie for Grandma!" But when his mama said, "No, sonny, this piece of pie is for you," he exclaimed with equal perturbation, "Oh, what a little bit!" His conviction was strongly biased by the point of view. "It makes a difference whose ox is gored."

The agnostic scientists claim, in substance, that there is so much relativity cumbered up with the presumption of absolute knowledge that we cannot be perfectly sure of really knowing anything, not even that we are sure that we can be sure of knowing nothing—that nothing that we know is absolute knowledge, and that knowledge is nothing that we know that we know.

However, most of us believe that there are a few things of which we can be reasonably sure, and one of them is, that things which are equal to the same thing are necessarily equal to each other, and a fact which will square with another fact will also square with a third fact which squares with that other fact. In brief, "What is sauce for the goose, is sauce for the gander."

Now the communities upon the shores of Lake Hopatcong are also important servants of the public. They, too, are public service institutions and they are rendering a far greater service by affording the people a highly attractive and healthful resort than could, by any possibility, ever be rendered by a rehabilitated and enlarged Morris Canal as a coal-carrier.

By consequence, then, the same reason for safeguarding the welfare of the public by safeguarding the interests of public service institutions also apply to the communities about Lake Hopatcong.

They have the same rights to protection of their interests under the law that any public service institution has, and those rights can not be circumscribed, limited or encroached upon in any way whatsoever without rank injustice, any more than could the rights of any public service institution, except when evidently larger public policy demands the exercise of the arbitrariness of eminent domain, with condemnation and due compensation, in order to work greater good for a greater number.

There is no other use which can be made of either the territory or the waters of Lake Hopatcong that will serve so many people and serve them so well as when utilized as a public park and summer and health resort.

Any disposition which may hereafter be made of the waters of Lake Hopatcong, either by the Canal Company, or by the Lehigh Valley Railroad Company as the lessee of the Canal, or by the State, when the Canal shall have become the property of the State, must, in justice, be subject to and must in every way respect all of the rights of the communities and dwellers upon the shores of the Lake, whether acquired by purchase or acquired by user.

When the State shall have acquired the Canal, no new rights will be created or can be created by that acquisition. The rights of the State will be the same as are the present rights of the Canal Company, or the Lehigh Valley Railroad Company as the lessee of the Canal.

The relation of the State or of the Federal Government to the individual, with respect to property rights, is exactly the same as the relation between individuals, or between corporations and individuals. One of the self-evident, eternal, cornerstone principles of justice and equity, upon which our Federal Constitution is founded, is the recognition and observance of the rights of every citizen to the absolute ownership and enjoyment of property as the main essential of liberty and the pursuit of happiness. Governmental recognition of individual property rights is the main bond which holds the States of the Union together, and it is the main bond of brotherhood of the people of the States, which makes them patriots.

In all ages the strength of a nation and a people has mainly depended upon governmental respect for individual property rights —the inviolate ownership of home and home lands and the sanctuary of the home in all that the word comprises and means.

When Rome was strong, her strength lay in the homes of her people. When Rome was invincible, individual property rights of the common people were inviolable.

No government is stronger than is the attachment of the people for the ordinations of that government which safeguard and protect them. In a government of the people, for the people and by the people, the interests of every individual are the concern of all. This makes the concerns of the government truly the concern of the people.

## The Morris Canal Should Be Abandoned

The question as to whether the Canal should be abandoned or rehabilitated and operated by the State, if dissociated from sentimental considerations, resolves itself into a question of prospective profit and loss.

Obviously, sentimental considerations could be sufficiently, and more rationally served, by making the Canal and its right of way into a cross-State parkway and driveway, rather than by rehabilitating and operating it at great loss merely to perpetuate it as a reminder of antiquated industrial enterprise.

If it is proposed that the State should take over, rehabilitate and operate the Canal purely as a business enterprise for profit, entirely aside from sentiment, then it behooves us seriously to inquire as to whether or not the operation of the Canal can, under modern conditions, be made a profitable instead of a losing business enterprise. If we learn through unimpeachable evidence that the State could not rehabilitate the Canal and operate it at a profit, but, on the contrary, must necessarily do so at great loss, then we should be convinced by this learning that the Canal should not be rehabilitated and operated by the State, but, on the contrary, that it should be abandoned.

Most of the uses for which the Morris Canal was built to serve no longer exist.

The charter of the Morris Canal was passed in 1824. At that time there was not a single railroad in the United States.

The first railroad built in the United States was a three-mile line from Quincy, Massachusetts, to tidewater, used for transporting granite from the quarries; begun in 1826 and completed in 1827.

The Morris Canal was begun in 1825, and completed in 1831.

The Morris Canal was not built as a competitor of railroads, for the railroad at the time of its construction was still in the experimental stage. To most serious-minded persons of that time, steam power was still largely a fangle of rumor and the imagination. The Morris Canal was built to compete with the wagon-train, and pack-mule team; and it proved a very successful competitor of them in the old pre-railroad days.

The Canal was well utilized for many years in the transportation of iron and zinc ores, with lime and charcoal for their smelting. The forests along the route of the Canal were cut and burned into charcoal for the smelters, sawed into lumber, or cut into cordwood, to be shipped by Canal to tidewater. But the lands were finally denuded of their forests; and ultimately the discovery of the great Lake Superior iron deposits rendered most of the iron mines of New Jersey unprofitable and worthless. The New Jersey lumberman is now a character of a bygone generation.

At the present time the Canal has scarcely any excuse for being except for the transportation of anthracite coal; and conditions have so changed since the coming of the railroads that transportation of anthracite coal by mule-power over the narrow, shallow, winding Morris Canal, with its numerous locks and inclined planes, uphill and down, can no longer compete with railroad transportation.

It must be remembered that the Morris Canal is of very exceptional construction. It is, unlike most other Canals, a real mountain-climber. It was the builders of the Morris Canal who introduced the system of inclined planes for mountain climbing, the boats being drawn up long steep slopes in wheeled cradles running on rails.

Every canal boat that passes over the Morris Canal either way has to be raised and lowered through an aggregate distance of about two thousand feet, so that all freight transported over the Morris Canal has to be lifted and lowered an aggregate distance of more than a third of a mile, for the height above the Canal to the top of the planes must be included.

Transportation over the Morris Canal is not all by water by any means. More than five miles is by rail over the inclined planes, and a very primitive, slow and inefficient sort of railroading it is.

As water must be used for generating the power for drawing the canal-boats up the long inclined planes and lowering them down again, together with the water consumed in operating the many locks, it is obvious that a far greater quantity of water is necessary for the operation of the Morris Canal than is necessary to operate canals running as is most usual, along practically level ground, while the loss by seepage and leakage is enormously greater from canals winding round sidehills than from those running over level plains.

The largest boats now run upon the Canal can carry about sixty-eight tons of anthracite coal. The crew of the Canal boat usually consists of captain, helper and two mules. It requires five days to make the trip from Phillipsburg to Jersey City.

The wages of the captain and helper and the fodder for the mules are now certainly four times as high as they were during the first forty years of transportation over the Morris Canal.

High wages and high cost of living make present day hand-labor expensive. Consequently, as the canal-boats are both loaded and unloaded by hand, the handling of their cargoes is much more expensive than is the handling of the cargoes of freight cars, which is done by machinery.

This is an age of machinery. The railroad represents the transportation and handling of coal by machinery, where the quantities are so great and the time and cost of handling and transportation so small compared with the time and expense of hand-labor involved, in transportation by Canal, that the problem reduces itself to one of competition between hand-labor and machinery.

Just as hand-labor with the sickle and the scythe can not compete with the mowing machine and the reaper; just as the hand-guided plow drawn by the ox-team can not compete with the steam gang-plow; just as the oar-driven barge and the sailing ship can not compete with the great modern steamship; so snail-paced Canal transportation with handling of freight by mule-power and hand-power can not compete with the steam-drawn freight train and with steam-operated machinery in the handling of cargoes.

But there is another and insurmountable obstacle to the rehabilitation and operation of the Morris Canal, and that is the utter insufficiency of available water to feed it. There never was water enough for the Canal in the days of its greatest usefulness. But there was much more water available then than there is today. The forests have been cut away, and tilled fields and barren lands have taken the place of woodlands. The rain is no longer held, as it once was, in the tree-shaded humus, moss and tangle of the forest, but runs off quickly from the unprotected fields, or soon evaporates in the sun, so that brooklets that were streams of considerable size sixty years ago during the months of July, August and September, are now entirely dry during those months.

The question does not resolve itself by any means into simply one of cost of transportation, even if transportation by the Morris Canal would be cheaper than by railroad, which it would not. There is the far greater question—adequacy of transportation.

## Utter Insignificance of the Morris Canal as a Coal Carrier

Even were the Canal to be rehabilitated and operated to its full capacity, it would still be quite an insignificant factor in supplying the demand for anthracite coal in New York and neighboring cities.

The proposition to rehabilitate and operate the Morris Canal as a benefaction to the users of coal anywhere in the State, is a scheme which, for sheer economic unwisdom, is a wonder.

The scheme would necessitate draining not only all of the watersheds once drawn upon to operate it, but also every gallon of available water that could be made to flow into the Canal would have to be utilized for its operation, and if it were to be enlarged as proposed, it would have to be shut down about half of the open season every year for lack of water.

The maximum freight carried by the Morris Canal in any one year, namely in 1866, was 889,220 tons. Coal constituted about half of its tonnage.

If the Canal were to be rehabilitated and operated, there would be practically no other freight except coal for it to carry, and consequently the freight would be all carried one way.

It is probable that 500,000 tons of coal could be brought by the Canal from Phillipsburg to tidewater in a year, provided the feed-water should hold out. Some persons who are not aware of the relation this quantity of coal bears to the requirements of coal consumption, may think that 500,000 tons is a very large quantity. As a matter of fact, however, it is an exceedingly small quantity.

Few persons have any conception whatsoever of the enormous quantity of coal that is consumed per annum in New York and surrounding cities, and few persons have any adequate idea of what enormous quantities of coal the railroads are carrying. It will surprise a good many to know that the Pennsylvania Railroad, for example, carries in three days as much coal as the Morris Canal could transport in a year.

The Pennsylvania Railroad carried 66,000,000 tons in 1910—one hundred and thirty-two times as much as could be carried by the Morris Canal, or it carried that year as much as the Morris Canal would be able to carry in a hundred and thirty-two years. The Lackawanna Railroad carried during that year about 11,000,000 tons—more than twenty times as much as could be carried by the Morris Canal. The different railroads entering New York and the near cities of New Jersey carried in 1910, the following quantities of coal:

| | Tons |
|---|---|
| Pennsylvania Railroad | 66,000,000 |
| Baltimore and Ohio Railroad | 37,000,000 |
| Philadelphia and Reading Railroad | 24,000,000 |
| Erie Railroad | 19,000,000 |
| Delaware and Hudson Railroad | 17,000,000 |
| Lehigh Valley Railroad | 14,000,000 |
| Delaware, Lackawanna & Western Railroad | 11,000,000 |
| Central Railroad of New Jersey | 9,000,000 |
| New York, Ontario and Western Railroad | 2,000,000 |
| Total | 199,000,000 |

Thus, the combined gross coal tonnage of the railroads with which the Morris Canal would have to compete is about 200,000,000, as against 500,000 tons for the Morris Canal. In other words, the Morris Canal would be able to transport only one-four-hundredth part as much coal as the railroads are carrying.

The coal which the Lackawanna Railroad alone carries in a year would make about 150,000 canal-boat loads, and it would require twenty canals the size of the Morris Canal to transport this quantity of coal; and yet the Lackawanna Railroad carries only five per cent. of the total quantity of coal transported by the railroads delivering coal to New York and surrounding cities. It would require four hundred canals like the Morris Canal to transport the quantity of coal which the above railroads carry.

The Lackawanna Railroad could transport from Scranton to Jersey City in 220 average trains of average cars as much coal as the rehabilitated Morris Canal could carry in a year.

Professor Moulton in his book, "Waterways Versus Railways," has shown that a two-track railroad, if devoted entirely to the carrying of coal, as the Morris Canal would be, would be able to transport in a year a hundred and five million tons of coal—two hundred and ten times as much as the Morris Canal could carry.

If all the railroads which bring coal to New York and neighboring cities of New Jersey were to carry nothing but coal, as the Morris Canal would do, they would be able to bring to these cities more than two thousand million tons (two billion tons) yearly. This is four thousand times the quantity of coal that the Morris Canal would be able to carry.

Consequently, as the Morris Canal would be in direct competition with all of the railroads supplying New York and neighboring cities with coal, even if coal were to be carried over it without cost and delivered at tidewater at the price at which it could be bought at the mines, it would tend to lower prices only one-fortieth of one per cent.

Were the railroads to handle only the same proportion of coal to other freight that they now carry, and the Morris Canal were to bring coal from the mines to tidewater free of cost, its influence to lower prices would even then be less than one-quarter of one per cent.

As it would actually cost very much more to carry coal by the Morris Canal than by railroad, higher prices would have to be charged for the coal it carried than would be charged by the railroads, unless the State of New Jersey should be generous enough to carry the coal at a loss as a benefaction to coal consumers, in which case the benefaction would extend equally to the inhabitants of New York State as to those of New Jersey, and the people of New Jersey would be bearing an onerous burden in order to lower the price of coal but a very small fraction of one per cent. mainly to New York consumers.

In manufacturing cities like Newark, Jersey City, Elizabeth, Hoboken, and Paterson, five tons per capita of the population per annum, is a low estimate for the quantity of coal consumed. Newark has nearly 400,000 inhabitants, and consequently consumes about 2,000,000 tons of coal per annum, so that the Morris Canal, if rehabilitated and operated, would be able to supply only one-fourth of the coal required by Newark, it would supply only one-third of the coal consumed by Jersey City, and less than four-fifths of the coal consumed by the City of Paterson.

Therefore, in order to supply the city of Paterson alone, through the Morris Canal, with about four-fifths of the coal it consumes, it would be necessary to draw upon an area about half the size of the State of Rhode Island to get the water, and the largest and most beautiful lakes of New Jersey would have to be drained to the last drop that could be drawn every summer to get feed-water to operate the Canal.

It must not be supposed for a moment that cheapened transportation of whatever sort would be able to lower the price of coal in any of the cities of the State of New Jersey, without simultaneously lowering the price in New York and all surrounding cities to correspond. Even if a railroad like the Lackawanna should bring coal from the mines to Jersey City free of charge, working up to its full capacity, any influence to lower the market price of coal would be as effective in New York City as in Jersey City. Prices, like water, find their level.

## Extravagant Waste of Water

If the Canal were to be rehabilitated and operated and every drop of available water utilized in its operation, the amount of freight which could be shipped over it would never exceed the maximum tonnage that used to be shipped over it in its palmy days, for there would not be sufficient water to feed the Canal.

On the contrary, the tonnage could never be made to equal that of fifty years ago before the lands were denuded of their forests, with their water-retaining humus, and before the drying up in summer and early autumn of springs and streams.

Not only are lakes and reservoirs necessary to store water for the Canal, but also large forest areas are necessary.

It is a conservative estimate that at least one-third of the water which used to be utilized to feed the Canal was stored and conserved in the forests of that period, whence it seeped gradually into the lakes, springs, and streams, for their maintenance during July, August, and September.

It has been proposed to enlarge the Canal, in order that boats of greater tonnage may pass through it. This expedient, instead of helping matters, would only serve to make bad matters worse, on account of the utter inadequacy of available water. It is a most significant fact that there never was sufficient water for the operation of the Canal in the days when its feed-water was more plentiful than now. Sometimes, when traffic was heavy, its feed reservoirs were exhausted and traffic had to be suspended.

If the Canal were to be enlarged, still more water would be required for its operation, and as far less water is available for the purpose now than formerly, its feed reservoirs would be very much sooner exhausted, and traffic would have to be suspended for a much longer period.

It is probable that if the Canal were to be rehabilitated and enlarged, it could not be successfully operated for lack of water from the middle of June to the middle of October, thus reducing by one-half the period when it would be open for navigation; and, in order to supply the water for the four months' navigation of the Morris Canal to transport not more than 500,000 tons of coal from Phillipsburg to tidewater—a quantity which the Lackawanna Railroad or the Lehigh Valley Railroad could transport in a couple of weeks, and the Pennsylvania Railroad in less than three days—New Jersey's most beautiful lakes would be depleted during the months when the water is most needed in them, and instead of their continuance as health-giving, prosperity-making summer resorts, they would be reduced to miasmatic lagoons and swamps for the spreading of mosquitoes and contagion.

Cranberry Lake, Greenwood Lake, and beautiful, splendid Lake Hopatcong, that now blossom with health and teem with life and joy and laughter through all the good old summer time, would, with depopulated and desolated shores, lie reeking in the sun, stinking to heaven for vengeance.

Northern New Jersey's most cherished rivulets and lakes would be ruined by the diversion of their water from its God-intended uses, in order to feed a navigable, seeping cesspool, five feet deep. forty feet wide, and a hundred miles long.

This simile of a cesspool is a good and pertinent one. The Canal's greediness for water to supply its seepage may be brought home to the country house-owner who knows what a quantity of water can be run into a comparatively small cesspool without overflowing it.

A cesspool five feet deep and ten feet across is considered an adequate receptacle for taking care of the waste water of a good-sized country house.

Imagine then, what a quantity of water is necessary to keep full to the brim a cesspool five feet deep, twenty-five feet across at the bottom, and forty feet across at the top; and yet the Morris Canal is a cesspool of such width and such depth and with a length of more than a hundred miles! Think of the stupendous quantity of water necessary to keep that cesspool full for purposes of navigation in spite of its tremendous seepage.

As I have pointed out elsewhere, C. C. Vermeule has shown that the seepage and leakage of the Morris Canal amounts to a loss of 1.74 cubic feet of water per second per mile, and as the Canal is more than a hundred miles long, the total loss throughout its entire length amounts to more than 174 cubic feet per second.

This would require a quantity of water more than double that which could be supplied by Cranberry Lake, Greenwood Lake and Lake Hopatcong combined. It would require more than five such lakes as Lake Hopatcong to supply the water lost by seepage and leakage from the Morris Canal.

Mr. Ford Kurtz, Civil and Hydraulic Engineer, who was employed by parties interested in the suggested electrification of the Morris Canal, and who could consequently, hardly be thought to over-estimate the loss from the Canal by seepage and leakage, but rather, if anything, to minimize that loss, has estimated it, in his Report, at 1.33 cubic feet per second per mile, which would be 133 cubic feet per second for the whole length of the Canal. Still, the loss, according to this minimum estimate, is three-quarters as great as estimated by that exceedingly competent and impartial engineer, Mr. C. C. Vermeule.

Mr. Kurtz very naïvely says:

> "My conclusions, as given in the report which follows, are 'that if the Morris Canal were put in a state of good

repair with full depth of water at all points and an additional net reservoired area of 10.4 square miles similar to present reservoired areas be secured, it would be possible even in the driest series of years to be expected to maintain navigation for a traffic of 1,000,000 long tons per annum with an average haul of 62.75 miles; and that the entire canal could be electrified from the flow required for navigation, without the diversion of additional water.' "

Therefore, in order to render Mr. Kurtz' plan for the electrification of the Canal feasible, it would require even additional reservoirs for impounding the feed water; and all of this water, in addition to the water drawn from all the other feed-water sources of the Canal, it is proposed to employ for the transportation of a million tons of freight per annum, on an average haul of 62.75 miles—a little more than half the length of the Canal.

Now, the value of such a quantity of water may conservatively be estimated at at least $5,000,000; and the value of lake and summer and health resort properties which would have to be sacrificed would surely be worth at least another $5,000,000.

The interest on $10,000,000 at five per cent. is $500,000 a year, and this, at the start, would impose an expense of at least fifty cents a ton for every ton of freight that passed over the Canal, even if a million tons per annum should be carried, which would be an utter impossibility from lack of water.

If the Canal should be rehabilitated and operated, the maximum quantity of coal which could be transported over it by the entire quantity of water available from present feed water sources, could not exceed 500,000 tons per annum, for it would be freight all going one way and the empty canal boats would have to return through the Canal to reload, so that the cost for water alone would amount to $1.00 per ton.

Mr. Kurtz very prudently makes the a priori announcement in his Report that the questions of the cost of the proposed electrification and the probable revenues to be derived from the improved Canal are not included within the scope of his Report.

The present rates for the transportation of coal by the Lackawanna Railroad from Scranton to Hoboken are as follows:

| | |
|---|---|
| Egg, Stove and Chestnut Coal | 1.58 per gross ton. |
| Pea Coal | 1.43 " " " |
| Buckwheat Coal | 1.28 " " " |
| Rice and Barley Coal | 1.13 " " " |

The largest boat which can be run over the Canal can carry only sixty-eight tons, and with all conditions favorable it takes five days to make the run from Phillipsburg to Hoboken, Jersey City, or Newark.

It must be borne in mind, however, that the Canal-boats must be returned again to Phillipsburg for re-loading, and as there is no freight to be carried on the return trip, they have to be taken back empty, requiring another five days, making ten days in all for the round trip.

The wages of Canal-boat captain and helper, with feed for the mules, for the ten-day trip to tidewater and return, added to the cost of loading and unloading the Canal-boats by hand, added to the cost of a sufficient toll to the State to maintain the Canal in a practical operative condition, and to pay the worth of the water consumed for purposes of navigation, and also to pay the interest on the investment, plus the wages of lock men and other attendants and operatives stationed along the Canal, and plus the wear and tear and depreciation of Canal-boats—all of which items must be taken into account in getting at the cost of Canal transportation—certainly make the cost of transportation by Canal far exceed the cost by rail, even though the Canal were rehabilitated and placed in the best possible condition.

The Lackawanna dumps its coal directly from the car, off trestles, or into wharf boats. The coal car is run on to a platform, and car and platform are hoisted and the coal is run out over the side of the car into a chute which conducts it into the boat.

The average harbor-boat or wharf-boat holds from 6,000 to 8,000 tons, while the largest Canal-boat holds only sixty-eight tons.

This forces a new consideration upon our attention, and it is the wharf and river space and basin space which would be required for the transportation and handling of the large number of Canal-boats necessary to supply any large coal-dealer or large consumer.

If their coal were to be supplied from Canal-boats, the boats would have to be towed about the river and bay by tugs at much expense, seriously interfering with other business upon the water.

There are many dealers in New York who put out as much as a thousand tons of coal per day at retail from their yards. This would require fifteen canal-boat loads a day; while an average dealer would require about ten canal-boat loads a day.

The Brooklyn Rapid Transit Company uses about a thousand tons of coal a day.

The New York Steam Power Company uses six hundred tons of coal a day.

The sugar and oil refineries use from five hundred to a thousand tons of coal a day.

The Edison Company now uses soft coal, but formerly this Company used 2,000 tons a day of anthracite coal.

According to Mr. Ford Kurtz' Report, 10,000 boats would pass through the Canal toward New York in a year, each boat capable of carrying sixty-five tons of coal, aggregating 650,000 tons. This would be 1,780 tons a day—not quite enough coal to supply two ordinary dealers in New York, and not quite enough to supply one concern like the Edison Company.

Now, if the Canal should be rehabilitated, it would be open to the use of anyone desiring to ship coal to any quarter. No partiality would be shown for the cities on the Jersey side. Consequently, it does seem rather a peculiar proposition to rehabilitate and operate the Morris Canal, making a Saharan desert and wilderness of a large part of northern New Jersey, with the destruction and depopulation of its most beautiful lakes and mountain summer resorts, in order to supply a single New York company with coal for carrying on its business.

If the State of New Jersey should take over and operate the Morris Canal in order to transport coal in competition with the railroads, it is a pertinent question how the Canal is to be supplied with coal at Phillipsburg, its western terminus, for the Morris Canal ends at Phillipsburg on the eastern side of the Delaware River.

It is true that the Lehigh Canal runs from the coal mines to Easton, Pennsylvania, on the western side of the Delaware River, but there is at the present time no way of transferring canal-boats across the river from one canal to the other, for the dam which provided this means was carried away by flood in 1894, and has never been rebuilt.

If the State of New Jersey should have the right to rebuild this dam and re-establish connections with the Lehigh Canal, it would still be necessary for the State to enter into a toll agreement with the Lehigh Coal and Navigation Company to work in conjunction with the Morris Canal.

In 1868, the Lehigh Coal and Navigation Company leased the Delaware Canal, which has since that time been operated in conjunction with the Lehigh Canal, and it would be impolitic and impracticable for that Company now to allow their Canal to be operated in conjunction with the Morris Canal.

Evidently the coal would have to be brought from the mines to Phillipsburg, over either the Lehigh Valley Railroad or the Central Railroad of New Jersey, and there transferred to the Morris Canal.

Now, as these two railroads are among the most important with which the State would have to compete, it would be the height of absurdity to bring the coal by these roads a third of the distance to Jersey City, then to transfer it mainly by hand into canal-boats, abandoning steam-power and towing it by mule-power around a tortuons, meandering, shallow, narrow, mountain-climbing canal a distance twenty-five miles greater than by rail, to Jersey City.

It must be borne in mind also that the total length of the planes on the Morris Canal is more than five miles, so that five miles of the distance would still be by rail, and a very poor sort of railroading it is, too, being a sort of compromise between a railroad and a hydraulic elevator.

The canal-boats have to be run into a cradle and drawn with a wire rope and windlass up long side-hills, and by the same means lowered again in the descent to Jersey City, the total distance which the canal-boats have to be lifted and lowered being more than two thousand feet—more than a third of a mile.

It would be very much cheaper and better for the State to hire the railroads to bring the coal right straight through to Jersey City rather than to transfer it to the Morris Canal.

## Summary of Reasons Why the Morris Canal Should Be Abandoned

1. If the Morris Canal were to be rehabilitated and operated, it would require the expenditure of more than a dollar's worth of water for every ton of freight that could be carried through it, and the value of the water necessary to feed the Canal would be far greater for other uses than for canal purposes.

2. There never was available water enough for the continuous navigation of the Canal, and now, owing to the small forested areas, there is not more than two-thirds as much available water as formerly.

3. If the Canal were to be rehabilitated and operated, the most beautiful lake and mountain resorts of New Jersey would be ruined in order to supply the feed water.

4. The loss of water by seepage and leakage from the Canal is more than double the quantity that could be supplied by Greenwood Lake, Cranberry Lake, Green Pond, Bear Pond, Stanhope Reservoir and Lake Hopatcong combined, the loss being 174 cubic feet per second, while the combined flow of these lakes is only 77 cubic feet per second.

5. The Morris Canal, unlike most other canals, is a mountain climber. Freight passing through the Canal in either direction must be raised and lowered through a distance of about two thousand feet—more than a third of a mile—by means of locks and planes. Consequently, travel is made correspondingly slow, tedious and expensive, and an enormous quantity of water in addition to what would be required for a more level canal is needed to operate the locks and to furnish the power to run the planes.

6. If the Canal were to be rehabilitated and operated, there would be no available freight worth considering except coal, and this would be freight carried all one way, the Canal boats returning empty, and the maximum annual tonnage of the Canal could not be made to exceed 500,000 tons on account of the shallowness and narrowness of the Canal and the small size of the Canal boats and the time necessary for the operation of locks and planes.

7. The distance by the roundabout way of the Morris Canal to Jersey City is about twenty-five per cent. greater than by rail.

8. The Morris Canal is able to bring coal only about two-thirds of the way from the mines to Jersey City, and the coal must be brought by rail from the mines to Phillipsburg and transferred largely by hand-labor to the Canal, there being no all-water route from the mines to Jersey City by way of the Morris Canal, for the Lehigh Coal and Navigation Company's canal ends on one side of

the Delaware River and the Morris Canal begins on the other side, with no way of bringing the coal across.

9. If the Morris Canal were to be considerably enlarged as well as rehabilitated, of course larger boats could be carried through it, but the annual tonnage would not be increased for lack of feed water, because the larger canal-boats would use up the water more quickly and correspondingly fewer boats would be able to pass through it.

10. If the Morris Canal were to be rehabilitated and improved in any manner, freight could never be carried by it as cheaply as it can be carried by rail. Consequently, the canal could never compete with the railroads in the cost of carrying coal, for the reasons already given and for the additional reason that the amount of hand-labor involved in carrying coal by the Canal as against the handling of coal with machinery when carried by the railroads renders the expense of transportation by the Canal prohibitive.

11. The quantity of coal, namely 500,000 tons, which could be carried per annum by the Morris Canal, is so very small compared with the quantities which the competing railroads are able to carry, that the Canal could never, under any circumstances, become more than a very insignificant factor in the supply of coal to New York and neighboring cities of New Jersey.

12. The quantity of coal which could be carried by the Canal, as compared with what can be carried by the railroads, is so small that even if the State were to operate the Canal to its full capacity, transporting coal absolutely free to Jersey City, it would have no real influence at all toward lowering the prices of coal, for the reason that the Canal could not supply one quarter of one per cent. of the coal which the railroads are carrying. In short, the Canal would be able to supply only one quarter of the coal required by Newark alone, and even less than would be required by Paterson alone.

13. Even were it true, which it is not, that the Morris Canal would be able to bring coal enough and cheaply enough to lower prices in Jersey City, Newark, Elizabeth, Hoboken and Paterson, those cities would not be able to reap any material benefit, for the reason that New York and other out-of-State cities would also reap equal benefit, for there could be no discrimination in favor of the cities of New Jersey.

14. The right of way of the Morris Canal would be worth infinitely more as a park driveway and portions of it for a trolley line for the accommodation of the public at large than it would as a Canal, for the Canal would be able to render but little service and that service one which would affect but a very small fraction of the population.

# What Should Be Done With the Right of Way of the Morris Canal

If the Morris Canal is abandoned, the question becomes a very pertinent one as to what disposition should be made of the Canal properties, especially of the right of way of the Canal across the State.

We all know, because the assumption is a self-evident one, that the best use which could be made of the Canal properties would be a public use which would best serve the greatest number of the people of the State. But how to dispose of the Canal properties so that the best interests of the people at large will be served is a problem to tease the wisdom of a Solomon.

Reason must, of necessity, proceed from the simple to the complex. Consequently, in approaching a complex proposition or problem, it is always wise to proceed as far as we can by simple rational processes based upon our co-ordinations of acquired experience and our sense of the eternal fitness of things.

Such methods of reasoning lead me to believe that the interests of the communities through which the Canal passes should be given especial consideration, and that, other things being equal, the disposition of the Canal within the limits of Newark, Jersey City, Paterson, Phillipsburg and other municipalities should be largely dictated by those municipalities.

Of course, it may be argued with reason that those municipalities would be actuated by local self-interest. Still, on the other hand, their local self-interest would naturally lead them to a more thorough investigation and understanding of the questions relating to their respective municipalities than persons outside of those municipalities would be likely to take the trouble of finding out.

Similarly, the best interests of the country districts through which the Canal passes should be carefully weighed, and such disposition should be made of the right of way of the Canal as would best serve those country districts; due and careful consideration, all the while, respecting both city and country sections, being given to the interests and welfare of the people of the State at large; and nothing should be conceded, granted or done, which would be inimical to the interests of the people of the State at large.

I think I have shown that if such a disposition should be made of the Canal as is desired and recommended by the communities around Lake Hopatcong, that it would serve the best interests of the people of the State at large.

The suggestion to convert a large part of the Morris Canal into a public parkway or driveway, appears worthy of serious consideration. In case this plan were to be carried out, the Canal should be filled up most of the way, and the wider spaces could be converted into beautiful parks. Along the route there would doubtless be numerous public-spirited persons who would gladly donate contiguous parcels of land for park purposes, for the reason that such parks along such a driveway would materially enhance the value of adjacent property.

Certain portions of the Canal right-of-way could also doubtless be well utilized for a trolley line.

It would not be wise, however, to leave the Canal open and maintain it full of water as a continuous park water-way. Such a measure would be exceedingly unwise.

We have shown what an extravagant waste of water would result from the rehabilitation and operation of the Morris Canal, in order to keep it full for navigation and to supply the water for operating the locks and planes.

It is a fact which few people appreciate, that only about five per cent. of the water required for the navigation of the Canal is expended in operating the locks and planes. Most of the water is required to supply the enormous loss by seepage into the ground.

The loss by evaporation, leakage and seepage combined, is more than twenty times as great as that needed to operate the locks and planes.

Consequently, if the open Canal were to be maintained as a parkway throughout its entire length, almost as much water would be required to keep it full as would be needed to feed and operate the Canal if it were to be rehabilitated. It would require more than double the quantity of water which could be supplied by Greenwood Lake, Cranberry Lake, Bear Pond, Stanhope Reservoir, Green Lake and Lake Hopatcong combined, in order to supply the feed-water to keep full such a parkway canal. Consequently, practically the same watershed area would be required for such a dead canal as would be required for a live and operating Canal. An area nearly half the size of the State of Rhode Island would have to be drawn upon for its feed-water. Lake Hopatcong and many other of the most beautiful lakes of New Jersey would have to be drained and converted into mud flats and mosquito incubators, in order to supply the feed-water.

The Canal should be filled up the greater part of its length, leaving an occasional pond or stretch of canal along the route where the character of the ground should be favorable, and where there would be a natural and constant supply of water flowing into and through the Canal to maintain circulation and prevent stagnation.

## A Plea for the Preservation of Lake Hopatcong. Speech of Hudson Maxim Before the Senate Committee on Railroads and Canals, at Trenton, New Jersey, February 5, 1912

Mr. Chairman and Gentlemen of the Committee:

Being a large property owner on Lake Hopatcong I am naturally deeply interested in any legislation to draw or divert the waters of the Lake for any new use, which is likely, injuriously, to affect the value of my property.

While speaking in my own interest and in defence of my property, I am speaking in the interest of all other property owners, and in defence of all other properties about the Lake.

I own about six hundred acres of land there, having about three miles of shore front upon the Lake.

Before becoming so deeply interested in Lake Hopatcong property, I looked carefully into the water question, and satisfied myself on several points; first that the Morris Canal and Banking Company had the right to draw down the waters of the Lake as might be required for the purpose of navigating the Canal, but that the Canal Company did not have the right to draw or divert the water for any other purpose whatsoever.

Furthermore, I learned, or thought I learned, that in the event of the abandonment of the Canal, which was expected soon, all rights to divert or draw the waters of the Lake would cease.

The laws of the State on which I have depended to safeguard my property, and the information upon which I have depended have likewise been depended upon by all other property owners, with the result that Lake Hopatcong has become a well-populated summer resort, and about its shores there are not only numerous summer cottages, but also many large hotels and palatial residences.

We, the property owners upon the Lake, have invested our money there under the best legal counsel and advice obtainable in this State.

We, the property owners, have been and are informed that neither the State nor any corporation has the right or can, without condemnation and compensation, acquire the right to draw or divert the waters of the Lake for any purpose whatsoever except to navigate the Canal.

In 1903, three Commissioners, Hon. George T. Werts, Hon. John W. Griggs and Hon. Foster M. Voorhees, all ex-Governors, and one of them at one time Attorney-General of the United States, were appointed under a concurrent resolution, to investigate and report upon the abandonment of the Morris Canal. Ex-Governor Voorhees of that Commission is now Chairman of the Water Supply Commission of this State.

It was the decision of these Commissioners (see Clause 6, page 37, of their Report) that

> ". . . . the water supplied to the Morris Canal, at the present time, and under present legislation, is water for the use or navigation of the Canal, and for no other use or purpose. The waters of the fresh water lakes and streams that constitute the supply of the Morris Canal, are private waters, owned by private parties. The State has no right or title thereto and can obtain the same only through purchase or by condemnation proceedings and making compensation to the private owners. To enable the Morris Canal to be used as a means for gathering and supplying potable water to the municipalities or inhabitants of the State, would require the enactment of legislation for that purpose. Under existing laws, neither the canal, nor any of the waters thereof, can be used for such purpose."

Also, see Clause 7, page 38, from which I quote the following:

> "To enable the State to own, or control, the water supply of the Morris Canal, would require a large expenditure of money to acquire such ownership or control through purchase or condemnation proceedings. We have no means of computing the extent of such expenditure, but it would undoubtedly be very large."

Before I became so largely interested in Lake Hopatcong property, I was shown the brief of Richard Boardman, counselor-at-law for Nathaniel Niles and Others, 1903, from which I quote the following propositions:

> "That the title to the soil covered by the waters of Lake Hopatcong is in private owners, and not in the Morris Canal and Banking Company, nor in the State of New Jersey.
>
> "A servitude in favor of the Canal Company to flow said lands for the purposes of its canal, but for no other purpose; to take from the Lake as much water as is necessary for the navigation of its Canal, but no more; and to apply the waters so taken to the purposes of its Canal, but to no other purpose.
>
> "That the servitude in favor of the Canal Company had its origin in the grant of the Legislature, expressed in the charter of the company.
>
> "That the servitude of the Canal Company is defined and limited by the terms of its charter.
>
> "That the servitude upon our estate of the Canal Company is a purely legal right, without equities of any

kind, not a dollar having ever been paid for it to us or to our grantors.

"That the right of flowage for other purposes than the navigation of the Canal is property vested in the owners of the soil under the Constitution, which cannot be taken from them except upon just compensation being made to them.

"That the Legislature cannot authorize the Canal Company to take more water from Lake Hopatcong than is needed for the purpose of navigating its Canal, except upon just compensation being made to the owners of the Lake (and to the riparian owners along the Musconetcong River).

"That the Legislature cannot authorize the Canal Company to apply any of the waters of Lake Hopatcong to any other than canal purposes, except upon just compensation being made to the owners of Lake Hopatcong (and to the riparian owners along the Musconetcong River)."

We laymen, in relying upon the laws of the State, are obliged to rely upon the advice of eminent lawyers and legislators for the true interpretation of these laws.

We have been advised that upon the abandonment of the Morris Canal, not one inch of water can be drawn from the Lake thereafter for any purpose whatsoever, and that the Lake cannot be made into a drinking water reservoir with the imposition of any new restrictions because of that use, without condemnation proceedings, with due compensation to all property owners.

We can often better understand a complex problem by looking at it in a simpler form. Suppose, for example, that I had a farm with a little lake of water in it, on the shores of which my residence and my farm-houses were built, and suppose that someone should come along and bargain with me, and secure the right at a nominal sum, to use the lake as a skating rink in winter, but for no other purpose whatsoever, and suppose the skating rink business should not pay, then, the question is, if the lessee should sell out his skating rink right to an ice company, would the ice company have the right to use the rink house for an ice house and cut and store ice from the lake; and would the ice company also have the right to prevent me from boating upon the lake and from bathing and fishing in it; and the right also to make me move my residence and farm buildings or raze them to the ground, to prevent contamination of the lake and injury to the ice, and all without compensation to me?

We recognize the fact that there have been and are certain disadvantages in drawing down the water of the Lake to navigate the Canal, and we are prepared and willing to bear the burden of these disadvantages, for we have always had and still have the freedom of the Lake for bathing, fishing and boating, and we have been able to build upon the shores, using cesspools for sewage disposal.

It is of far less concern to us, because of far less disadvantage to us, that the waters should be drawn down, as has been done to operate the Canal, than that the Lake should be made a drinking water reservoir, with the prohibition of boating, bathing, fishing and building upon the shores, and prohibition of cesspool disposition of sewage.

Even if the Lake should then be kept full, the advantage of continuous high water would not equal the disadvantage of the restrictions which would inevitably follow the making of the Lake into a drinking water reservoir. Obviously, without the right to boat, bathe and fish, and without the right to reside upon the shores of the Lake, it would make but little difference to the property owners about the Lake whether or not the water were to be drawn down much or little; for without the freedom of the Lake all the shore properties would become practically valueless.

We do not pretend to deny the right of the State to take the waters of the Lake for any purpose whatsoever, by condemnation proceedings, but we do maintain that they cannot be taken for any new use, which, by consequence of the nature of that new use, would be disastrous to Lake properties, without due compensation to the owners of those properties; and if any bill is passed the terms of which violate our legal rights, we believe, and are advised that we may safely rely upon our belief, that such a bill would be unconstitutional, and that any provision which, by its terms, should infract our rights, could not be enforced.

If we are correct in our understanding of this matter—an understanding acquired by us of the best dispensers of legal wisdom in the State, then, whatever rights we have under the laws of the State are actual, and not hypothetical, and are rights which cannot be ignored.

As citizens of this State, our rights in the State ought to be respected. It is to the best interest of all other citizens of the State that our rights under its laws be respected, for the welfare of each citizen under the law's protection is the concern of all.

It has been suggested that a compromise be effected, whereby the Lake may be drawn down from three to four feet for potable purposes, and that no onerous restrictions be placed upon the use of the Lake as a summer and pleasure resort, with boating, fishing, bathing and building upon its shores, and it has been urged that the State will not want the water for potable purposes for at least twenty years hence, and that, in the meantime, we shall have a full Lake.

There are several serious troubles with such an arrangement. In the first place, what is to happen twenty years hence to the water of the Lake is effective now upon property values. If anything is to happen then which, if it should happen now would be injurious, the injury to property values is effective now.

Again, why should we compromise? All of the rights to the waters of Lake Hopatcong which we granted to the Canal Company, either do or do not revert to us upon the abandonment of the Canal, and these rights either are, or they are not, wholly ours absolutely; and if they are, why should we compromise? And if they are not ours, then why should we be compromised with? Why should we be considered at all?

Again, after a water main has been laid into the Lake, how could we prevent more water being drawn than had been agreed upon?

In this connection, we must take into consideration this fact, that in dry seasons, when there would be the greatest need to draw the waters of the Lake down to the limit of the compromise, the Lake, without drawing, would by evaporation and seepage be lowered to that limit, and there would be no water available for potable purposes.

Again, if it should be agreed now that no onerous restrictions shall be imposed later, how shall we be able to prevent their imposition at any later time, if we cannot prevent it now?

Let us look at this compromise problem in the simple form of a simile: Suppose an acquaintance comes to me and demands that, whereas I have a hundred dollars in my pocket, I should compromise with him and give him fifty of it, holding out the inducement to me that by letting him have the fifty I should prevent him taking the whole, as he would be willing to compromise on that sum.

Those of you gentlemen who have visited Lake Hopatcong, know the significance of the truth that it is the most beautiful summer resort section of this State, and that it is destined soon to become one of the most important of all-the-year-round residential sections.

I have spent more than a hundred thousand dollars to develop and improve my property at the Lake during the past seven years, and am this year building a road through my property at a cost of more than six thousand dollars, which will connect with other roads that are in course of construction, so that soon there will be an automobile driveway clear around the Lake, and this driveway is sure to become one of the most attractive and popular short tours in the country.

Do you know, gentlemen, that Lake Hopatcong is a large body of water, about seven miles long, with a shore line, including the islands, of about forty miles in length, and that this Lake is almost a thousand feet above the sea, and that mosquitoes are about as rare there as are snakes in Ireland?

Do you know of the very large number of passengers carried to and from Lake Hopatcong every year by the Lackawanna Railroad and the Central Railroad of New Jersey? And do you know that it is already one of the most popular points visited by automobilists, and that it is one of the main attractions that brings automobilists into the State?

It has been reliably estimated that each automobile party visiting or touring the State spends on the average about fifteen dollars a day.

At the present time Lake Hopatcong brings into the State many thousands of automobiles every year, and it is a lodestone that makes them linger longer than they otherwise would, and leave their money here, and the profits to the State on the money they leave is even now greater than the interest would be on that sum of money at which the Lake Hopatcong water rights are appraised.

At the present time, Lake Hopatcong is, furthermore, an important tax-paying community; the property there is an important asset to the State. But we are not to think merely about the present moment in considering a section of the country like Lake Hopatcong, where building is going on so rapidly, and where property values are doubling every few years.

We must consider also what that community will be and what its significance to the State will be ten years from now—twenty years from now—generations from now!

Again, we must well weigh the significance of the proximity of Lake Hopatcong to New York City—the metropolis of the western world—and the fast growth of rapid transit facilities. All the tunnels under the Hudson, later the bridges that are to be built over the Hudson, and all the buildings that go up in the City of New York, serve to enrich all of New Jersey.

If Lake Hopatcong be dedicated as a permanent residential section and summer resort, millions of dollars will be diverted in their flow and brought into this State by the attractions of Lake Hopatcong, and large and expensive villas will be built in Northern New Jersey, which would otherwise be erected upon the shores of the Hudson or on other waters; for the city man in summer is bound to have his mountains, his lake or river waters, and the cool airs, bright suns and skies that go with them.

Think what Niagara Falls means to the State of New York! The torrential flow of its mighty waters brings into New York State a concomitant torrential flow of mighty dollars.

If the waters of Lake Hopatcong are to be diverted to flush city sewers, and that resort is to be wiped off the map, then there will flow away from that Lake and away from this State many times more money than all its waters are worth for potable purposes.

Lastly, every citizen in the State of New Jersey is proud of Hopatcong. Every one who has ever visited it is enthusiastic about Hopatcong; and thousands of New Jersey citizens enjoy its beauties and are renewed in health there every year.

With the example of New York State's esthetic devotion to Niagara in preserving that waterfall to gladden its citizens and the world with its beauty, when millions could be reaped by selling for power purposes all that the power companies are anxious to buy, cannot we, the people of New Jersey, well afford to preserve Lake Hopatcong, and shall we not preserve it, for all the people's greatest good?

## Speech of Hudson Maxim Before the Morris Canal Investigation Committee at the Lake Hopatcong Hearing, November 16, 1912

Mr. Chairman and Gentlemen of the Committee:

Lake Hopatcong, queen of all the lakes of New Jersey, is on trial for her life, and if she is to be sentenced to destruction, it means an actual sentence of death also upon numerous persons in this and neighboring States who would otherwise be afforded opportunity of finding salvage here, because the numbers of those who have already found at this lake the parting of the ways from fatal disease to sound health are beyond our counting.

Every resident of Lake Hopatcong here tonight is recalling to his mind at this moment many such cases, and there are some among those present who would now be but a memory had it not been for the life-saving grace of Lake Hopatcong.

Three years ago, a certain literary gentleman, who is now our acting Borough Clerk, and one of us this evening, was a pale, emaciated consumptive, in the pent airs and disease-laden whirl of New York City, and one day he fell in the street stricken with that premonitor of death, the pulmonary hemorrhage. He was taken to a hospital, and the attending physician counted his remaining days upon one hand, but a friend brought him to Hopatcong, and in less than six months he was well again, and fat and sound, and not a germ of tuberculosis to be found by the most searching diagnosis. He is today a perfect picture of health and manly vigor.

The son of the proprietor of this hotel, who is now such a rosy, vigorous, handsome lad that he is a feast for the eyesight and we turn and look at him a second and a third time for the mere gladness of it, was a puny, pale and sallow despair four years ago when he came to Hopatcong.

The disposition of Lake Hopatcong, if and when the Canal is abandoned, involves two main considerations: First, actual, legal vested rights and interests of the residents and property-owners upon the shores of the Lake and in adjacent communities; and, second, the dominant rights and interests of the general public—of the people at large.

Law is an attempt to govern future procedure in the light of past practice. At best, all jurisprudential procedure is but an approximation to the natural adjudication of the human conscience, based upon our sense of the eternal fitness of things.

A commonwealth is a corporate body which, like any business corporation, fulfils its true purpose insomuch, and only insomuch as it is a truly co-operative body, where all citizens are shareholders, just as truly as are the stockholders in any great business corporation. By consequence, individual welfare is part and parcel of the general welfare, and individual concern is part and parcel of public concern just in proportion as individual personal value and the value of individual possessions are elements of and contributory to the greatest good of the whole corporate body of the people.

Condemnation of property under the right of eminent domain, is a species of surgical operation where some limb of the body politic is sacrificed for the salvation, or best interest, of the rest of the body as a whole; and, just as is true with the human body—the more vital a member and the more it is contributory to the life and welfare of the whole organism the more is it valued and safeguarded and reluctantly sacrificed—so with the body politic.

Those members which contribute more to the general health, general prosperity and general good, should not be sacrificed to a cause or use less contributory to the general welfare, and the sacrifice should not be made of one part or member in the interest merely of another part or member, but for the whole body.

What is the best use to which Lake Hopatcong can be put? Is it worth as much for a water reservoir as it is worth for a summer and health resort?

In spite of the ever-present Canal problem, many properties around the Lake have doubled in value several times during the past twenty-five years.

If, as Mr. Morris Sherrerd has told us, the water would not be needed for potable purposes for the next twenty-five or thirty years, perhaps longer, and if, as we have every reason to believe, population will continue to increase and property values around the Lake to enhance in the same ratio for the next twenty-five years, then there will be a population at this resort of 100,000 persons, and there will be $100,000,000 worth of property. These are conservative figures. And it is such a population and such a resort that has been proposed to be sacrificed in order that, twenty-five years hence, the water of the Lake may be drawn away to supply a population no greater than Lake Hopatcong's population will be at that time.

On Page 142, Vol. 3, of the Geological Survey of the State of New Jersey, of 1894—"A Report on Water Supply," by C. C. Vermeule—the present storage capacity of Lake Hopatcong within the whole flowage basin formed by the eleven foot raise of the Lake by the Canal Company, is estimated at 1,100,000,000 cubic feet. This amounts approximately to 7,000,000,000 gallons.

This quantity of water would supply Newark's 400,000 inhabitants for only 175 days, or less than six months. This quantity of water would supply 175,000 inhabitants for one year. Last winter at Trenton it was suggested to compromise on drawing the water down four feet instead of eleven feet. If this were done, then the water available would supply only 64,000 inhabitants for one year, or the whole city of Newark for but sixty days, and the Lake would be sacrificed for that.

On Page 194 of the same volume, we learn that the loss of water from the Canal by seepage and leakage is about 1.74 cubic feet per second per mile. Now, as the Canal is more than a hundred miles long, the total loss for the entire Canal, exclusive of water lost by evaporation and exclusive of water used for lockage and for operating the planes, is about 174 cubic feet per second.

Therefore, the 1,100,000,000 cubic feet of water which Lake Hopatcong holds would supply this loss for only about seventy-three days, so that it would take five Lake Hopatcongs to supply the water lost by leakage and seepage from the Canal.

Now this matter of loss by leakage and seepage is something to be taken into account by the advocates of the Canal Parkway proposition.

If it is better and more to the interest of the people of the State at large that Lake Hopatcong should be dedicated as a contributory supply of potable water for certain municipalities, then it should be dedicated to that use, but it may as well be understood from the first, and the people at large should understand from the first the whole true significance of its dedication as a potable water supply. The people at large should understand from the first that such a dedication would mean the prohibition of all boating, bathing and fishing upon the Lake and the total exclusion of all persons from the Lake.

The majority of the people of the State should understand and appreciate from the first that they would be surrendering this beautiful summer and health resort without *quid pro quo*—without compensation.

The people at large should understand that they would be making a donation at great self-sacrifice by cutting themselves out of the right to participate in the enjoyment of this Lake by dedicating it as a potable water supply.

This State has an estimated population at the present time of about 3,000,000 inhabitants. All of these three millions would be robbed and irreparably injured in order that about five per cent. of their number should draw the waters of this beautiful Lake away to flush their sewers, sprinkle their streets and irrigate their gardens, and, incidentally to get a few bucketfuls for kitchen use.

It is perfectly right and proper that the interest of the few should be subverted, with due compensation, in the interest of the many. That is lawful and that is best, and lawful because best, which serves the greatest good of the greatest number. The subordination of the interest of the few to the welfare of the many is called the exercise of the right of eminent domain, but not the subordination of the rights of the many to the interests of the few.

The 3,000,000 inhabitants of the State of New Jersey all have certain legal rights to visit and enjoy Lake Hopatcong with their friends and families. If the Lake be dedicated as a State Park, the people will retain their own for their enjoyment. If, on the contrary, it is dedicated as a potable water supply, it can, as I have shown, serve but 175,000 persons, or only about five per cent. of the population of the State.

It would be a most extraordinary, most unjust and most ridiculous procedure, and an inversion of every principle of the right of eminent domain, to sacrifice the rights in Lake Hopatcong of ninety-five per cent of the population of the State in the interest of less than five per cent. of the population.

It is the opinion of eminent engineers that in the near future every municipality of size will have a double system of water supply—a low pressure system for potable purposes, and a high pressure system for the extinguishment of fires and for commercial uses—and that with the growth of population in rural districts, the use of surface waters accumulated in lakes and rivers when used for potable purposes, will have to be purified, and with the two-fold plan of municipal water supply, where only the potable water will need purification, this will not be a hardship.

Lake Hopatcong's value as a potable water reservoir can not exceed what it would cost to construct a reservoir elsewhere for the storage of an equal quantity of water.

A reservoir forty-four feet deep would require to have only one-quarter the surface area of Lake Hopatcong, in order to hold as much as Hopatcong, namely, 1,100,000,000 cubic feet. In other words, it would require a reservoir less than a square mile in area.

In order to hold a quantity of water equal to that obtainable by drawing the Lake down four feet, it would require a reservoir only one-third of a square mile in area and forty-four feet deep. If made twenty-two feet deep, it would have to be only two-thirds of a square mile in area. If made as shallow as fourteen and two-thirds feet deep, it would require to be only one square mile in area.

The value of a lake or waterfall, or any other piece of Nature, cannot always be appraised on the basis of its actual economic worth, for often a cluster of hill, water and sky possesses a far greater than economic value even though the economic value be immense.

Perhaps Niagara Falls affords as good an exemplification of this truth as any piece of natural scenic beauty. The economic value of Niagara as a water power is so enormous as to transcend our comprehension.

The customary method of appraising a water power is to calculate the sum of money which, placed at interest, would yield a sufficient income to buy the coal necessary to develop steam-engine power equal in amount to the water-power.

Now it has been estimated (See United Editors' Encyclopedia, Vol. 26; also "Hydroelectric Developments and Engineering," by Frank Koester, Page 327), that Niagara Falls develops an energy equal to 16,000,000 horse-power. To develop 16,000,000 horse-power through the best up-to-date steam engines would require the consumption of about two pounds of bituminous coal per horse-power hour. Consequently, it would require 32,000,000 pounds of bituminous coal per hour, or 16,000 tons per hour.

This is a quantity of coal so great as to represent a stream of compact coal 3½ feet square running through a chute at a speed of 16 feet per second continuously forever, piling up 384,000 tons every day.

At $3.00 a ton laid down at the plant, this daily quantity of coal would cost $1,152,000. Approximately, then, it would cost a million dollars a day to buy the coal necessary to develop the steam power equivalent to the energy developed by Niagara Falls. Deducting Sundays and holidays, the amount would be approximately $300,000,000 a year—a sum which in gold would weigh 450 tons, a sum which to raise would require a tax of three dollars a year on every inhabitant of the United States.

When coal is burned or any water tumbles down a cataract, the energy developed is the release of stored-up sunlight.

To the purely economic mind, Niagara Falls is equivalent to a cataract of coal a foot thick and twelve feet wide, falling over the brink at the rate of sixteen feet a second. To the merely utilitarian mind, Niagara Falls represents an economic waste for mere public sentiment and entertainment comparable to Nero's burning of Rome; in fact, a sum so vast as to provide for public amusement a conflagration of the size of Nero's bonfire every six months.

It is estimated that 13,300 buildings were destroyed in the great London fire, the property loss being estimated at about $50,000,000. Consequently, a conflagration the size of the great London fire could be provided once every sixty days for public entertainment and amusement for a sum equal to that which is running to waste at Niagara.

What then is the true value of Niagara, and what are we to take as the real coefficient of its value? It would require $6,000,000,000 placed at interest at 5% to yield $300,000,000 a year, the sum necessary to buy the coal to develop steam power equal to the water power that is annually wasted there.

This sum is equal to four-fifths of the total wealth of the Empire of Japan, and yet the majority of the people of the State of New York and of the United States, and of the Province of Ontario and all of Canada appraise the value of Niagara for its unequalled scenic beauty and sublime grandeur so highly as far to outweigh its economic value as a water power, and effective measures have been taken to restrict its use for power purposes.

This greatest wonder of the world is a world-posssession, and all the people of the world are interested in its preservation for its aesthetic and greater than economic value. To give over Niagara to the vandalism of private enterprise would be a crime against nature, a violation of the right of eminent domain of the people of the whole world.

The Palisades of the Hudson—those grand old towering sentinels—tell of a long bygone pre-glacial age when the Hudson was a river so mighty that the Hudson of today, grand as it is, is but a rivulet in comparison! There is a lot of excellent trap rock in the Palisades. It is most admirably located for quarry and transportation, but the hand of the people is raised to stay the hand of the vandal, and the Palisades are to be preserved.

Central Park cost New York City $5,000,000 sixty years ago, and what a great tribute it is to the splendid foresight of the city forefathers. How inconceivable is its value today! How far above any economic worth is the value of that nearly a thousand acres of free earth, free air and open sky to the people of the great city!

During the glacial period an all-wise Providence, with ice and boulders, ground this large and beautiful Lake Hopatcong out of the earth and rock. It is New Jersey's greatest piece of scenic beauty. It is New Jersey's natural Niagara Park. Just as the State of New York dedicated Niagara Park to its people, so should the State of New Jersey dedicate Lake Hopatcong Park to its people.

## The State's Great Vacation Park

Now, as a matter of fact, not only are the property owners in and about Lake Hopatcong interested in the Lake, but also all the people of the State have interests and personal property rights in that large and most beautiful mountain Lake. The interests of all the people of the State are directly allied with the interests of the people of Hopatcong. The difference between their interests and ours at Lake Hopatcong is merely one of size or degree.

The inhabitants of Mars, being farther away, do not get so much of the sun's warmth as we do, still their interest in the sun is as actual as ours. Even the inhabitants of far-off Neptune, which frontiers the solar system, to whom our sun is only a star, are still interested as actually as we are in the warmth that they do get from the great central furnace.

Similarly, every inhabitant of the State of New Jersey has rights as actual in Lake Hopatcong as any property owner about its shores.

Lake Hopatcong is the common people's summer and health resort of the State of New Jersey—the one where the common people can go for recreation, rest and recuperation at a minimum of expense, to camp, to bungalow, or to cottage on its shores, or to board at some excellent boarding-house or hotel at moderate rates; to spend their vacation bathing in its waters, and in fishing and boating, and in roaming over the rock-cliffed hills and through the fragrant woods.

Had the Canal Abandonment Bill passed the Legislature last winter in the original form in which it was presented, Lake Hopatcong would have been utterly destroyed as a summer and health resort of the people, and instead of being a great crystal smiling in the sun—New Jersey's lodestone of health and happiness—it would have been doomed to conversion into a municipal, potable water supply tank, and during most of the good old summer time, it would, with depopulated shores, be reduced to a series of lagoons, bogs and swales, exhaling deadly miasmas.

The property owners of Lake Hopatcong made common cause with the interests of all the people of the State in trying to prevent the passage of that bill in such form as to doom Lake Hopatcong to ruination.

The water of Lake Hopatcong would be sufficient to supply for municipal uses only about five per cent. of the population of the State, and had that bill passed, the interests of the other ninety-five per cent. would, without any *quid pro quo* whatsoever, have been sacrificed to the five per cent. who would have got the water.

The blessings realizable from giving, above those realizable from receiving, may be sufficient to have rewarded the ninety-five per cent. of the inhabitants of the State, had they relinquished, without any equivalent, their rights in Lake Hopatcong, in order to make a present of its waters to five per cent. of their numbers.

I plead guilty to selfishness in wanting to protect my property at Lake Hopatcong. I also plead guilty to selfishness in wanting to protect the other property owners around the Lake, and I further plead guilty to having a selfish interest in serving the selfish interests of ninety-five per cent. of the population of the State.

The best servant of the people is he whose own individual interests are in common with those of the people whom he serves. The man who has no self-interest in common with the people he is serving never has been, never will be, and never can be so faithful a servant as he who, in serving the people, serves himself at the same time.

Not only does Lake Hopatcong directly serve the people of the State as a summer and health resort, but also it brings into the State from surrounding States a very large number of summer sojourners, many of whom ultimately buy country homes and settle down here.

Lake Hopatcong brings into the State during the summer and early autumn months a continuous procession of automobiles, bringing much business to the hotels, and bringing higher prices to the farmer for everything he produces and sells. Each automobile party spends, on the average, more than fifteen dollars a day.

Every farmer in the State must place to the credit of Lake Hopatcong a part of his yearly profits. Throughout all northern New Jersey, every farm is worth more dollars today and will in future be worth many more dollars still, because of that beautiful and prosperous summer resort—Lake Hopatcong.

Every farmer who desires to sell the whole or part of his lands stands a very much better chance of doing so and of getting good prices, in the northern part of the State, where Lake Hopatcong is, than he would if that beautiful sheet of water were to be made into a city water reservoir, and drawn down during the hot months of the year, to breed mosquitoes and disease.

Suppose it were possible for the Federal Congress to pass a bill which would destroy Niagara, in order to supply electrical power to the cities of New York State. It is self-evident that the people of all the other States would suffer a great loss and be done a great wrong in order that a few should have electrical power for their cities.

## PROPERTY MAP OF LAKE HOPATCONG

### Showing Its Wonderful Adaptability for a Summer and Health Resort and Its Equally Wonderful Lack of Adaptability for a Potable Water Supply

---

The attached map of Lake Hopatcong, which I have had prepared for the Morris Canal Investigation Committee, is the most accurate and complete that has yet been made of the Lake and its shore properties.

My surveyor, Mr. P. E. Boomer, C. E., spent more than four months in making special surveys, in gathering information and in the execution of the map.

The numerous red squares indicate the buildings that have been erected upon the shores of the Lake, and these, together with the numerous streets and roads indicated in the various communities, show to what a large extent this important summer resort has already been built up, and what a multitude of cottages, hotels and business houses are clustered about its shores, all of which would have to be sacrificed and this beautiful summer resort converted into a wilderness, if the Lake were to be taken as a drinking water supply for cities.

Not only must we take into account the significance of the sacrifice of the present population and property about the Lake, and the present significance of this beautiful summer resort to all the people of the State, but also, just as the sacrifice of a promising youth robs the community of the coming man, so the sacrifice of this growing summer resort, in its promising youth, would rob the State of the stupendous benefaction which its future would mean.

No city of the State today stands in need of the water of Lake Hopatcong. No one claims that the water of Lake Hopatcong would be needed during the next twenty years. To kill this prosperous summer resort, on a contingency that the water might possibly be needed twenty years hence, is the height of absurdity, for there are most ample sources of water supply from great and sparsely populated watersheds, where the sacrifice or cost would be small, and where the water would receive the minimum of pollution, and this in addition to the practically inexhaustible supplies of pure water obtainable by sinking artesian wells.

In ten years from now, the population of the State will be nearly four millions, and in twenty years from now, it is estimated that the population will be five and a half millions.

Even now, Lake Hopatcong would, as a water supply, benefit only about five per cent. of the population of the State; in ten years from now it would benefit about four per cent. of the population, and in twenty years from now, it would benefit only three per cent. of the population of the State. If drawn down four feet, according to the proposed compromise of last year, the water of Lake Hopatcong would even now benefit less than two per cent. of the population of the State.

In twenty years from now Lake Hopatcong will itself have a population as great as that which its water would supply for municipal purposes.

The very irregular form of Lake Hopatcong, winding and meandering as it does among the hills, making numerous bays, coves, and inlets, gives it an exceedingly long shore-line in proportion to its area.

Lake Hopatcong has a shore-line, including its islands, of about forty miles, while the area of its surface is less than four miles.

A round lake having the same area as Hopatcong would have a shore line of only 6.93 miles—only one-sixth as much shore-line as Lake Hopatcong has. Lake Hopatcong, with its less than four square miles of surface, has as much shore-line or lake front as would a round lake having a hundred and twenty-eight square miles of surface.

Consequently, Lake Hopatcong has as much available shore-front for the building of cottages as would a lake of ordinary round shape more than thirty times as large.

The very irregular contour of Lake Hopatcong, and its very long shore-line, most admirably adapt it for a summer resort, not only because of its wild beauty, but also because of its very great capacity for accommodating a summer population. But these very things which admirably adapt it for a summer resort also render it correspondingly unsuited for a potable water reservoir, unless the entire population around its shores should be removed and their property destroyed, which certainly would have to be done if any of the water were to be diverted for city uses.

Even were Lake Hopatcong to be destroyed as a summer and health resort, to make it a city water supply, it would still be a very impractical potable water reservoir, because, owing to its exceedingly irregular form and long shore-line, it would cost many times as much to patrol it and protect it from the intrusion of the public as it would to protect an ordinary potable water reservoir of the same area. In short, Lake Hopatcong is wonderfully adapted for a summer resort and equally wonderfully ill-adapted for a potable water reservoir.

# SOME Important FACTS About Lake Hopatcong

### Showing how Wonderfully it is Adapted for

## A Public Park, Summer and Health Resort

### Also, how Badly Adapted it is for

## A Source of Drinking Water for Cities

---

LAKE HOPATCONG has an area of less than four square miles.

Lake Hopatcong has a shore line of more than forty miles.

A round lake of the same area would have a shore line of less than seven miles.

There are a hundred and eleven lakes in the State of New Jersey, and Lake Hopatcong is the largest and most beautiful of them all, but Lake Hopatcong is situated on the top of a ridge of hills and has a watershed only one-twelfth as large in proportion to its size as is the average size of the watershed of the lakes of the State.

Consequently,

The water in Lake Hopatcong is changed only one-twelfth as often as is the water in the average lake, so that with an equal number of inhabitants on its shores in proportion to the quantity of water it holds, Lake Hopatcong would become twelve times as polluted as would the average lake in the State.

Lake Hopatcong can accommodate a shore population equal to what could be accommodated by a chain of six round lakes, each of them having an area equal to that of Lake Hopatcong.

Consequently,

Lake Hopatcong has advantages as a summer and health resort equal to a chain of six round lakes, each of them as large as Lake Hopatcong.

Lake Hopatcong would, with an equal density of shore-front population, be polluted seventy-two times as much as would a round lake of the same size having the average watershed; in other words, it would be polluted more in a week than would the round lake in a year.

Lake Hopatcong has as much shore front as a round lake thirty times as large—that is to say, a lake having 128 square miles of surface.

Consequently,

Lake Hopatcong can accommodate a shore front population equal to a round lake thirty times as large.

Consequently,

Lake Hopatcong would be exposed to thirty times as much pollution as would a round lake of the same length of shore line, other things being equal—but other things are not equal, on account of the small size of the watershed of Lake Hopatcong.

It is an amazing fact that the water of Lake Hopatcong, would, therefore, be subject to three hundred and sixty times as much concentration of pollution from shore-front population as would a round lake, having the same amount of shore-front and having the usual-sized watershed; in other words, Lake Hopatcong would be polluted as much in a day as would the round lake in a year.

Furthermore, the Lake Hopatcong watershed receives, per square mile of surface, much less rainfall than the surrounding country. This fact was noted by that eminent engineer, Mr. C. C. Vermeule, in his "Geological Survey of New Jersey," Vol. III., Water Supply, 1894, pages 13 and 103.

As an actual fact, the average annual rainfall for the State of New Jersey, taken over a long period of years, is 46.14 inches, while the average annual rainfall for the Lake Hopatcong watershed is only 42.54 inches, so that the Lake Hopatcong watershed gets 3.60 inches less rain in a year than does the average watershed in the State.

Now, 3.60 inches over the entire watershed represents a shortage, on the 3.82 square miles of surface of the lake, of two feet, so that Lake Hopatcong receives 213,000,000 cubic feet (1,500,000,000 gallons) less water than it would if it got the normal rainfall.

Spotted something that isn't quite right?

Let us know and we'll fix it for future reprints. And as a thank you for your diligence, we'll send you a free Sothis book of your choice.

publisher@sothis.press

Made in United States
Orlando, FL
05 November 2026

90231068R00028

This page intentionally left blank

Except of course for that line above. And this one.
And whilst we have your attention, perhaps you'd
like to follow us on Instagram @sothis.press. Go on.